MW00713789

To Kathleen –
Happy Bellowing!
Steve Swinburne
3/12/03

BOXING RABBITS, BELLOWING ALLIGATORS

COURTSHIP POEMS FROM THE ANIMAL WORLD

STEPHEN R. SWINBURNE

THE MILLBROOK PRESS
BROOKFIELD, CONNECTICUT

To every conservationist working to safeguard the habitat of boxing rabbits, bellowing alligators, and other life on earth.

—S.R.S

Cover photograph courtesy of © Manfred Danegger/NHPA
Photographs courtesy of © E.A. Janes/NHPA: oo, 2,3,16,17; Animals Animals: pp. 5 (top © Gerard Lacz; bottom © E.R.Degginger), 6-7 (© Patti Murray), 8-9 (© Michael Fogden), 20 (© Richard Kolar), 23 (© Gerald Lacz), 24-25 (© R.Blythe /OSF), 26-27 (© E.R. Degginger); Photo Researchers, Inc.: p 10 (© Kenneth H. Thomas));Bruce Coleman, Inc.: pp 12-13 (© Mark Newman), 18-19 (© Kim Taylor); Peter Arnold, Inc.: pp. 14-15 (© Albert Visage); © Nigel Dennis/NHPA: p. 22; © VIREO: pp. 28 (A. Morris), 29 (J. Williams)

Library of Congress Cataloging-in-Publication Data
Swinburne, Stephen R.
Boxing rabbits, bellowing alligators: courtship poems from the animal world / Stephen R. Swinburne.
p. cm.
Summary: Depicts how the male in various species will call attention to himself in order to attract a mate.
ISBN 0-7613-2556-5 (lib. bdg.)
1. Courtship of animals—Juvenile literature. 2. Males—Juvenile literature. [1. Animals—Courtship.] I. Title.
QL761 .S85 2002 591.56'2—dc21 2001006485

Printed in the United States of America
5 4 3 2 1

Published by:
The Millbrook Press, Inc.
2 Old New Milford Road
Brookfield,Connecticut 06804
www.millbrookpress.com

CONTENTS

BOWERBIRD

Handsome bowerbird, charming bowerbird,
How do you meet a mate?
Make a stick house on the floor of the forest,
Decorate the front porch of your bower.
Display things blue,
Like a marble or toy.
Where a bowergirl meets a bowerboy.

LIZARD

Shy lizard, secret lizard,
How do you meet a mate?
Hug the sunniest spot on the trunk of a tree,
Nod your head like you never will stop.
Under your chin
You hide beautiful skin.
Flash it! Your fan is on fire!

BULLFROG

Chubby bullfrog, bigmouthed bullfrog,
How do you meet a mate?
Take a deep breath
From the marsh, misty night,
Sing like a deep bass drum.
Wake up the moon:
Jug o' rum, jug o' rum, jug o' rum!

PEACOCK

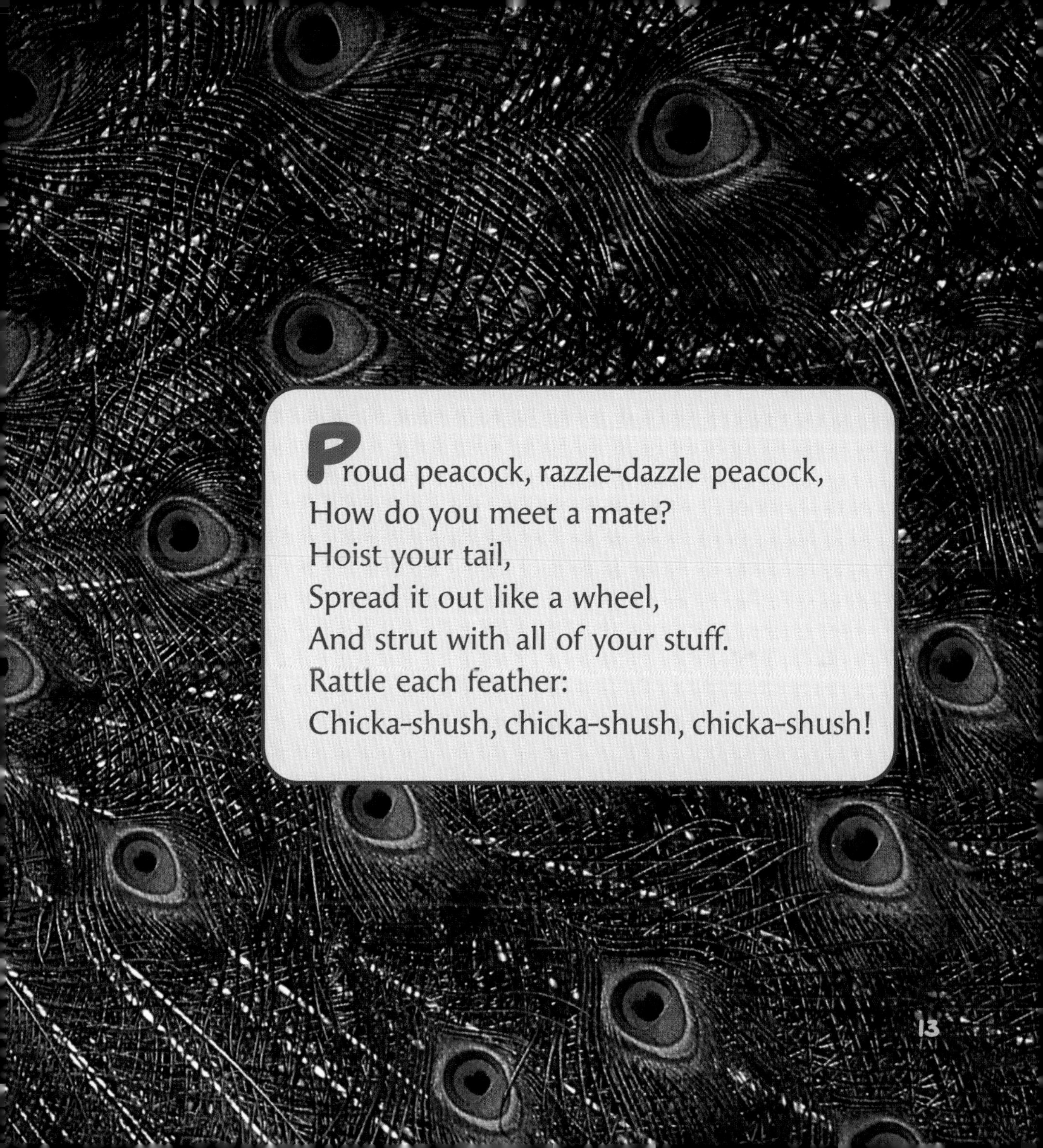

Proud peacock, razzle-dazzle peacock,
How do you meet a mate?
Hoist your tail,
Spread it out like a wheel,
And strut with all of your stuff.
Rattle each feather:
Chicka-shush, chicka-shush, chicka-shush!

ALLIGATOR

Sleepy-eyed alligator, scaly-backed alligator,
How do you meet a mate?
Puff up your throat,
Bellow like thunder,
Make the water sizzle like soda.
Belly-flop your head, splash!
Then smile that gator grin.

RABBIT

Long-legged rabbit,
lickety-split rabbit,
How do you meet a mate?
Stand tiptoe tall,

Put up your dukes,
Jab and swipe and cuff.
Watch out for the kick!
And may the best boxer win.

STICKLEBACK

Cool-colored stickleback,
zip-zooming stickleback,
How do you meet a mate?
Build a watery nest,
As round as a bubble.
With sticky threads,
To hold it together,
A wish come true for a lady fish.

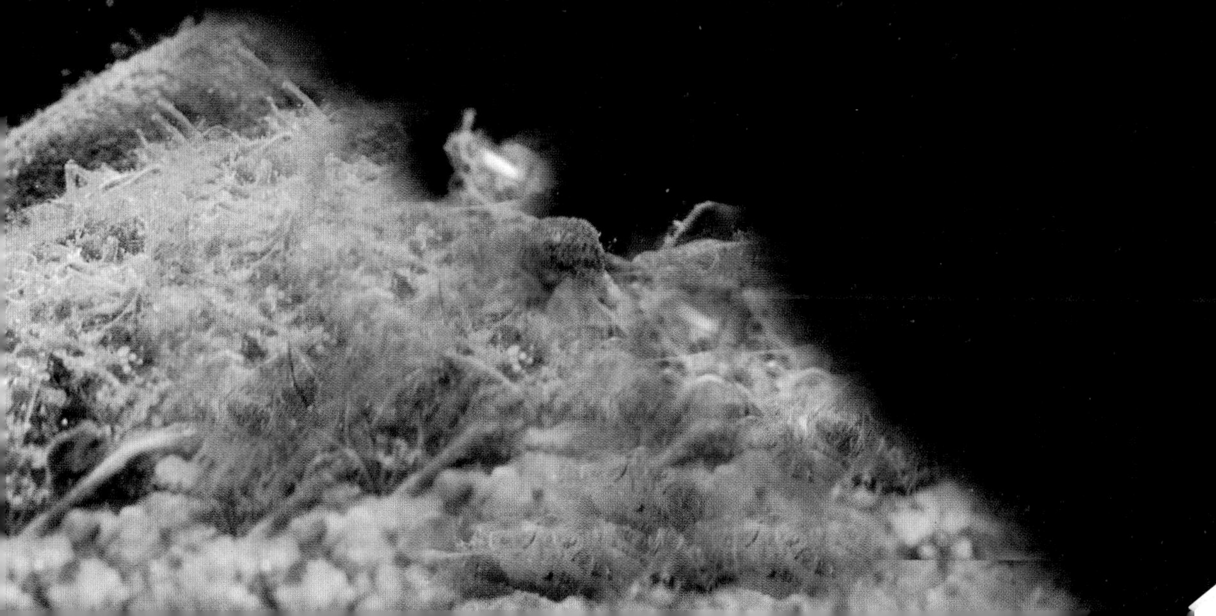

FRIGATEBIRD

Elegant frigatebird, sky-hugging frigatebird,
How do you meet a mate?
Throw back your head,
Point your beak to the sky,
Make your pouch as fat as the moon.
Waddle and wear
Your big, red balloon.

HIPPOPOTAMUS

Floating hippopotamus,
lazy hippopotamus,
How do you meet a mate?
Yawn wide like an open gate,
Show off a pair of square teeth,
Charge and stop,
Splash down, splash up.
Squirt water! Your nose is a hose.

FIREFLY

Friendly firefly, hide-and-seek firefly,
How do you meet a mate?
Flit and dash
High over the grass,
Sprinkle the night with light.
There's a torch in your tail!
See it wink, see it flash.

FIDDLER CRAB

Big-clawed fiddler, sand-dancing fiddler,
How do you meet a mate?
Hang out near your burrow,
Scan the sand with beady stalked eyes,
Wave fiddle to the sky,
Swing it low and take a bow,
And do it again and again and again.

WHOOPING CRANE

Tallest whooping crane,
rarest whooping crane,
How do you meet a mate?
Start the dance with a do-si-do,

Face your partner, bow real slow,
Flap your wings, leap sky-high,
Point your beak and bounce, bounce, bounce.
Put on your best bird ballet show.

ANIMAL FACTS

Satin bowerbird

The satin bowerbird lives in Australia. This coal-black bird is the size of a blue jay. The bowerbird has found a unique way of attracting a mate. The male builds an archway or bower made of little sticks on the forest floor. Then he looks for small things to decorate his house. Sometimes he steals them from a neighbor. Satin bowerbirds love blue—blue clothespins, blue straws, blue bottle tops. He makes an eye-catching house.

Lizard

Lizards, like snakes and turtles, are reptiles. They live in warm, tropical places. Lizards pick a favorite post or branch when it comes time to look for a mate. They bob up and down as if they were doing pushups. At the same time, they display a colorful patch of skin below their throat. This bright pink or yellow skin is called a dewlap. Like a colorful flag, the lizard's throat fan attracts attention.

Bullfrog

Bullfrogs are the largest frogs in North America. They live in ponds, lakes, and rivers. On summer nights, male bullfrogs sing in their deep, hoarse voices. Their call sounds like "better go round" or "jug o' rum." Bullfrogs do not sing in a chorus like spring peepers or toads. They sing solo and hope the right mate is listening to their best "jug o' rum."

Peacock

Peacocks are native to India and Sri Lanka (formerly Ceylon). You can often see them in zoos. The male peacock is one of the biggest showoffs in the bird world. When interested in meeting a mate, the peacock raises long, colorful feathers at the base of his tail. He makes a beautiful fan of feathers above his head. He parades around trying to catch the eye of his mate.

Alligator

Alligators are large reptiles that live in warm, tropical areas. They do not have bright colors to help them attract mates. The alligator has found another interesting way to get attention. The male alligator bellows very loud, mooing like a cow. At the same time, vibrations from the call makes the water frizzle or dance all around for a moment or two. At the end of his performance, the alligator slaps his head, making a big splash.

Rabbit

Male rabbits know how to get attention. A pair of rabbits will stand on their hind legs and box each other. Often these fights are nothing more than pushing contests. But sometimes the fight gets serious and one rabbit will kick another. Whoever wins the boxing match wins the mate.

Stickleback

Sticklebacks are small fish, only one to three inches long. Some live in freshwater and some live in the ocean. Sticklebacks are champion nest builders and that's how they attract a mate. Using his mouth, the male stickleback builds a hollow, round nest from stems, twigs and leaves. He makes doors at either end. The stickleback cements the nest together with sticky threads from his mouth. In his new fine house, his mate will lay eggs.

Frigatebird

Magnificent frigatebirds spend most of their lives flying over tropical oceans. When it's time to meet a mate, the male bird finds an island and builds a nest of twigs. Standing on his nest, the frigatebird sucks air into the wrinkled pouch under his throat. The pouch inflates into a big red balloon. The frigatebird drums the pouch with his beak to attract attention.

Hippopotamus

Hippos live in the rivers and lakes of Africa. They feed in short grasslands for a few hours at night and spend the rest of the time in water. To win a mate, male hippos begin with a yawning contest. They yawn and show their big front teeth. They rush forward and dive. They grunt and blow water through their nostrils. They splash up and down. The whole show takes place in the middle of a river.

Firefly

Fireflies or lightning bugs are common, small beetles. You can see their small points of light on dark, summer nights flashing and winking across a lawn or field. A male firefly uses the flashlight in his tail to search for a mate. He zigzags above the ground, his tail blinking off and on. When a female firefly watching from the grass sees his light, she blinks her tail, too. They keep flashing until they meet.

Fiddler Crab

Fiddlers are small crabs that live in the muddy or sandy banks of salt marshes. Male fiddler crabs have a large front claw. In summer, during the breeding season, hundreds and hundreds of males stand by their burrows and wave their large "fiddle" or claw to attract a mate. This big claw is always the right claw. If the claw is lost in battle, a new large one will grow at the next molting, only this time the big fiddle will be on the crab's left side.

Whooping crane

The whooping crane is the tallest bird in North America and one of the rarest. It breeds in the Northwest Territories of Canada and migrates through the Great Plains to winter in Texas. In their wild prairie home, male and female whooping cranes perform an amazing mating dance. Pairs mate for life. The whooper's alarm call of ker-loo! ker-lee-oo! can be heard for several miles.